To Alfonso Apdal Amos, his family and friends, and to all who have known and loved a silent soul.

Silent Soul

The Alfonso Apdal Amos Story

THE PRICE OF FREEDOM CHRONICLES:
COLD WAR ERA

Charleston, SC
www.PalmettoPublishing.com

Silent Soul
Copyright © 2023 by The Price of Freedom Foundation

The Price of Freedom Foundation
312 Christian Drive
White House, TN 37188

Cover design by Kristen Ingebretson

ISBN: 978-1-958969-00-7 (hardcover)
ISBN: 978-1-958969-01-4 (paperback)
ISBN: 978-1-958969-02-1 (ebook)

Contents

Foreword

As Founder of The Price of freedom Foundation I welcome you to our first published biography.

Our purpose is to collect, preserve for future generations, and tell to as many people as possible the life stories of those who paid the ultimate price for our nation's freedom – those who died while in military service. Our intent is to honor them, the lives they lived, the service they performed, the sacrifice they made, the price they paid for our freedom, and the price their surviving family and friends continue to pay.

This story, the story of Alfonso Apdal Amos, and those that will follow, helps survivors recall, and in some cases discover, elements of their lost loved one's life: who he was; what his hopes and dreams were; how he was in his various relationships with family and friends; as husband, father, friend, and more. It documents and preserves his story for future generations of his family and friends.

For those who did not know him, this book brings more than simply a face and name to one of the many thousands who lost their life in service to our nation. This work shows the very real man who served. The man who served his family and served our nation. His is a story both unique to him and bearing similarities to many who serve today.

Keith Hayden is our writer, and he has done an outstanding job on this work.

I was introduced to Keith by Jen Amos. Jen is the daughter of Alfonso Amos who first spoke with me in December of 2020 about her father and how our very new organization might help her and her siblings pull together the story of her father's life they had been working on for about a year at that point.

Keith and I met via a Zoom meeting and I was immediately impressed with him. His background as a military Veteran, and Air Force Academy graduate, as a criminal investigator for the Air Force, and as a published author all made me thank God for the introduction! But more than that, it was Keith's heart for our mission, telling the stories of those who are unable to tell them for themselves, that made me certain that he was perfect for the role we needed him to perform.

Prior to meeting Keith, we had recorded interviews with all the family and friends we could identify who knew Alfonso. The family provided photographs and many hours of home videos and videos of the memorial services the Navy provided for Alfonso and his family. We had requested military records. Keith was able to get the records of the investigation into the disappearance of MM1 Amos through a Freedom of Information Act (FOIA) request.

Keith put his heart and soul to work, as he will describe in the Preface, and produced an integrated story that I hope you find as compelling as I do.

Dennis E. Schroader Sr.
LTC, US Army retired
Founder of The Price of Freedom Foundation
www.PriceOfFreedomFoundation.org

Preface

An unexpected connection led me to write this story.

I met Alfonso's daughter Jennifer Amos after reading one of her husband's books. She ran a podcast called "Holding Down the Fort" where she had conversations with military spouses about their experiences running businesses and living life while living the military lifestyle. As a military spouse who runs his own business, she invited me on as a guest. During our talk she mentioned Dennis and the noble mission at The Price of Freedom Foundation for the first time. It was from him that I became aware of Petty Officer First Class Alfonso Amos' story.

I had never written a military biography before writing "Silent Soul". However, I had plenty of experience writing my fiction and drafting investigative reports as a former criminal investigator. This strange combination paired with my fascination with history along with my interest in telling stories for those who can't speak for themselves was the beginning of a 5-month long endeavor to piece Alfonso's story together in a manner that captivated readers as well as honored the legacy of his service in the United States Navy.

Shortly after meeting Dennis I set to work. It took months of digging through military records, reviewing interview records, and research, but the result was one of the most unique pieces of writing I've ever created. One you will not soon forget.

Part historical sojourn, biography, and record of service with narrative elements "Silent Soul" is a reminder some of the toughest battles don't take place in the air or at sea. It is a memorial informed by my respectful inquiry and love from Alfonso's friends and family. An epitaph functioning as a continuation of an ongoing tale of loss, mystery, and sacrifice told from my perspective as someone who wanted to understand his life and spread the word of his quiet professionalism as a sailor and father.

Let this living biography serve as proof of his life for Jen, the rest of her family, and those who wish to honor all departed warriors who bleed for strangers and nations at home and in distant lands. May it lead to your greater awareness of the untold stories of those who selflessly undertook the profession of arms, yet could never tell their story after years of wordless duty.

Keith
Practicing composite creativity
https://keithhayden.net/

Introduction

The mighty USS Kitty Hawk traveled rough chilly waters of the Pacific on the night Alfonso Apdal Amos disappeared.

Days before, Typhoon Zeb had formed in the Philippine Sea, picking up moisture, speed, and strength over the warm water's surface. In four days, it would swell to super typhoon status to become one of the largest storms of the 1998 typhoon season. With maximum sustained winds of 205 kilometers per hour (125 miles per hour) the storm reached its apex on October 14, 1998. The next day the biblical force of nature would continue its path and crash into Luzon, the Philippines's most populous island and location of Alfonso's birth.

Petty Officer Amos probably knew of the storm. As a career sailor with eighteen years of service in the United States Navy and at least two with the Philippine Army, he had seen countless storms in his life. The ship to which he was assigned, the USS Kitty Hawk, was no different. At nearly forty years of active service in the fleet, dodging, weathering, or cutting through storms was a common occurrence for, at the time, America's only forward-deployed aircraft carrier.

The storm—known as Typhoon Iliang in the Philippines—ripped through the islands, destroying homes and infrastructure while causing floods, landslides, and widespread crop destruction. Total damage estimates were around $126 million USD ($219 million in 2022), a disastrous figure.

After passing the Philippines, the massive storm weakened over Taiwan as it cut a boomerang path north toward Japan. By October 17, the USS Kitty Hawk Narrowly passed the gale force winds of the storm, riding the chop of the rolling sea as it reached the coordinates 25° 49.3N, 134° 13.2E. A location in the open ocean 303 kilometers (188 miles) from the nearest land and 1,213 kilometers (754 miles) from its homeport in Yokosuka, Japan. Alfonso's duty station at the time.

On the evening of October 17, Petty Officer Amos was scheduled to stand watch—a common duty for many enlisted sailors at sea—from 2000 to 0000 hours.

Unbeknownst to his fellow shipmates and his wife and three children back at Yokosuka, probably winding down for the evening; it would be his final watch.

CHAPTER 1

Son of a Farmer

*"As long as I am with him, I am already
happy because of who he is as a brother."*
- Celedonia (Shirley) Amos Guiang
(Alfonso's sister)

Alfonso Apdal Amos was born on December 19, 1954 in San Marcelino, a small town located in the province of Zambales in the Philippines. Located on the west side of Central Luzon, the nation's largest island, at the time of his birth the town was home to no more than fifteen thousand residents.

Photos of the area depict lush green mountains surrounding placid bodies of water. The town itself appears as a Spanish inspired small community with the wild jungle of the area growing in pockets around local churches, res- taurants, and busy city streets. But back in the 1950s, the area was much less developed and presented a more pastoral setting, with dirt roads linking the town center to the adjacent farmland scattered across the plains.

In the grainy footage of an old VHS home video, the person filming (assumed to be Alfon- so) captured the beauty of the area from a bus. As it rumbled across twisting dirt roads it over- looked a valley thick with vegetation. A small river snaked through the valley floor, provid- ing perspective of the considerable drop from the road. There were no guardrails to prevent vehicles from tumbling into the natural cradle. Alfonso spent most of his youth among this pic- turesque greenery, down the road from where he shot the footage.

Alfonso's father, Eulogio Abaya Amos, served as a city councilman for San Marcelino at some point during the 1960s. A position his only daughter, Celedonia (Shirley) Amos Guiang said, "was a sign everyone loved him."

The elder Amos was also a farmer. He intro- duced young Alfonso and his brother Rhoger to farming life early on. Shirley recalled with fondness Alfonso and his brother being summoned to the farm when they were all younger. If they did not obey, her father would whip them. In her memory, Alfonso always showed up when told. Rhoger got the whip.

The youngest of the siblings, Shirley had a close connection with her brother that extended into adulthood. She described him as a helpful, kind, and loving person, who made time for her even after he was deep into his military career, living a world away, with a family of his own.

When they were children, with no computers or internet to amuse themselves, they played in the old ways. Hide-and-seek was an old favorite, providing hours of joyful fun and excitement. As he grew up, Alfonso developed an affinity for basketball and farm work, yet remained helpful and protective toward his baby sister. Their relationship would endure the strains of his nomadic military career in the future.

Alfonso's mother, Dolores Apdal Amos, was a peaceful churchgoing woman. She was Eulogio Abaya Amos's second wife after his first was accidentally shot and killed. In the future,

Dolores would follow Alfonso to America to live under his care and help raise her grandchildren, but the nature of Alfonso's relationship with his mother is unclear. When asked about it by a medical professional during his military career, Alfonso described it as "good." The succinct response was characteristic of his quiet personality. He likely had many more thoughts and emotions stirring within him, but rarely expressed or documented them in any way.

Life in San Marcelino was bucolic and peaceful. The easygoing pace was evident in a home video shot by Alfonso on a return trip sometime in the early 1990s.

The camera fixes on an open courtyard, dusty and with a small well for pumping water by hand and a simple wood fire for cooking in the corner. The area is enclosed by unpainted concrete structures. The side of one small building is comprised of vertically aligned sturdy logs and sheets of metal. A buffer from annual storms like Typhoon Iliang. There is no excess here. Only the necessities of life and those treasured by Alfonso.

An earlier video, recorded in the late 1980s showed people gathered together, most likely to celebrate the rare presence of Alfonso and his family at home.

The setting is idyllic. In dry grass, with a single large tree present for shade, men smoked, women chatted casually in their native tongue, and kids played among the adults with green rolling hills in the background.

After some time, Alfonso hands the camera to someone else. He strolls over to a cooler, grabs a drink from it, then takes a sip, while silently taking in the relaxing atmosphere. An occasional smile runs across his face. He appears at peace.

Growing up on a farm, it was common for Alfonso to see chickens, goats, and other animals around his house. This connection to nature was reflected in one of his favorite hobbies: fishing.

Alfonso's love of fishing was legendary and a well-known fact among his peers and family members. In the local ponds and streams around San Marcelino, his boyhood diversion would transform into a lifelong passion and respite from the stresses of military and family life.

His son, John Paul Amos, recalled, "there were always four or five fishing poles around the house when I was growing up." His father's fishing pole collection was a testament to the enthusiasm he had for the activity.

That enthusiasm influenced the young John - or "JP" as his parents often called him when he was young. Fishing became a treasured activity for JP. A time when he could have his father all to himself. For Alfonso, it was an activity that grounded him in his simple roots, and his attempt to make up for time lost from his boy and family after lengthy periods of separation at sea.

———

We do not know the specific reasons why Alfonso decided to enter military service. However, it is apparent several factors may have influenced his decision.

The first was because of his father. Eulogio, who had spent time in the US as a younger man, had obtained a green card. Due to his experience in the States, he felt there would be greater financial opportunity if his sons went to America. As a result, he encouraged Alfonso and his brother, Rhoger, to cross the Pacific to begin a new life. Rhoger, eager to leave home, left first, but Alfonso hesitated.

According to his friends and family, farming life had taken a hold of him, and he had no desire to leave San Marcelino, much less travel to America. Rhoger commented, "He was scared to go."

We will never know the basis for Alfonso's fear. Like many of his other emotions he kept his true motive locked up in the vault of his psyche.

His decision not to leave the Philippines for the United States immediately may have factored into the second reason he decided to join the military.

There were few available ways to make money in the Philippines at that time. His sister Shirley commented, "The Philippines was not very developed at that time, so it was hard to find jobs."

The lack of opportunity combined with his father's wishes for him to have a stable life pushed Alfonso toward military service.

If he harbored any reservations to joining the military, he kept them to himself. Silent agreement, with possible strong opposition to his circumstances lurking beneath the surface unexpressed, would become a trait Alfonso was known for. One that would have dire consequences later in his life.

CHAPTER 2

Armed Rebels

"...to protect themselves...he was involved
[in combat], also."
- Marcelina Pamintuan Amos
(Widow of Alfonso)

Alfonso's decision to join the Philippine Army would set him on a path he would follow for the rest of his life. His introduction to military life would thrust him in the mix of an armed rebellion within his home country.

The Philippines of the 1970s was a chaotic place. Having achieved complete independence from the United States on July 4, 1946, the young nation struggled through a variety of economic, social, and internal military conflicts that would play out through that decade and beyond.

On November 11, 1969, Ferdinand Marcos was elected President of the Philippines for a second time. Marcos, a World War II veteran, with his slick black hair and unassuming features, looks the picture of equanimity and poise in many of his photos at the time. However, in stark contrast to his demeanor, he would become one of the most notorious dictators of the twentieth century. His draconian policies led to over 70,000 human rights abuse complaints being filed. Among these were 3,257 extrajudicial killings, 35,000 individual tortures, 737 disappearances, and 70,000 incarcerations.

In response to civil unrest, economic woes, and the growing shadow of communism, Marcos declared martial law in September 1972, sparking one of the darkest periods in the history of the country for the next nine years. Some of his most egregious abuses included press censorship, the torturing and killing of political opponents, and the imprisonment and assassination of outspoken critics.

It is unclear whether martial law or any of the other events during this period influenced Alfonso's life directly. But by 1970, at the age of sixteen, he most certainly saw the tumult of the events in local newspapers or heard whisperings of widespread dissent in the halls of his school.

Could the momentum of the times have propelled Alfonso into military service? Or was it the desire to follow his father's advice that caused him to act?

His motivations are unknown to us, but what is clear is that at some point in the early to mid 1970s, Alfonso joined the Philippine Army. A decision that marked the beginning of his life of military service for his home country and eventually in the United States.

To know the nature of the conflict Alfonso encountered during his time in the Philippine Army is to understand some of the history of the southern portions of the Philippines, the Sultanate of Sulu, and the native Moro people.

Formed on November 17, 1405, the Sultanate of Sulu was an Islamic kingdom founded by Sharif ul-Hashim, an Arab Muslim and religious scholar born in what is now the country of Malaysia. For hundreds of years the Sultanate ruled over a vast area, including the island of Borneo and parts of the present-day Philippines. By 1815 the Sultanate was recognized by Great Britain and Spain as a sovereign state.

During this time, the sultanate controlled eastern Sabah, northern Sarawak, and most of Mindanao, the southern island of the Philippines. Its reach extended far beyond its borders, to include all of Palawan. However, during this golden era, the nation had no standing army, nor were they ever recognized as a sovereign nation by the United States.

The Sultanate had maintained power for hundreds of years, but they eventually lost their autonomy to Spain and Great Britain in the mid-1800s, which brought about an era in which none considered them a governing body or an organized territory. After this time, through negotiations with America and the Commonwealth of the Philippines, they were officially recognized as a territory with limited governing power. The Sultan continued to maintain some control over that region. A half-hearted gesture that would leave many political, economic, and cultural issues unresolved and lead to continued armed resistance well into the mid-twentieth century.

A discussion of the source of conflict in the southern Philippines would be incomplete without mentioning the role of Islam and the Moro people.

The Moros are the inhabitants of Muslim Mindanao, an area with its own history and identity distinct from the rest of the archipelago. The word "Moro" originated from the Latin term "Mauru" and was the name the Spanish used for anyone who practiced Islam whom they encountered in the Philippines.

Tuan Masha'ika set foot on the Sulu archipelago in 1280 where he established the first Muslim community. Islam arrived in the southern Philippines in the late 1300s when trade routes were established between the local population and Melaka, a port settlement on the west side of the Malay peninsula.

Islam continued to spread throughout the region in the following centuries. By the time of Ferdinand Magellan's arrival in 1521, he found a strong Islamic monarchy with claims over most of the Sulu islands. As one researcher put it, "Eventually, it is the Islamic faith that would distinguish the Moros from the other natives of the Philippine archipelago."

1565 was the year that the Spanish entered Mindanao—also referred to "Moroland" by some—to impose Christian colonial dominance on the population. Using a divide and rule strategy to enlist the assistance of pockets of Filipinos already converted to Christianity, the Spanish effectively undermined Moro resistance for centuries, leading to occasional violent skirmishes and battles between both sides. The Spanish-Moro conflict, as it is commonly referred to in most historical texts, is considered to be the longest anti-colonial war in history by many historians, lasting for about 330 years.

The superlative "longest anti-colonial war" is a subject of debate among historians. Even though Moro resistance against the Spanish officially ended in 1898 after American annexation, their fight would continue in the form of guerilla resistance against the Americans, then eventually with the government of the Philippines. Evidence from a historical scholar further proves this point:

> "The southern Philippines, especially those regions that are inhabited by Indigenous or Muslim Filipinx [Moros] were forced to become part of the Philippines when the country earned its Independence in 1945-1946. The previous agreements between the colonized Philippines and the American government prior to World War II, was that the Muslim territories were independent of American colonialism as in they did not have to answer to the American government and puppet government put in place and that they could maintain their sovereignty. To avoid more fighting, and because the American military, similar to the Spanish, could not encroach and defeat these regions, it made a contract

with these southern regions of the
Philippines that they would not
fight one another [or], acknowl-
edge their independence. But
of course the United States lied
and never even to this day ac-
knowledges this agreement. So
when the Philippines became a
new nation after the world war,
they tried to take in the southern region, and a lot of this was
of course done by force. Because
the southern Filipinos continued
to use the previous agreement
as historical and legal proof that
they were a separate entity from
the Philippine nation, the new
Philippines government used
similar tactics learned from the
United States and labeled these
leaders and their communities as
rebels, communists, and terror-
ists. The issues over land rights
and independence became more
heated under Marcos and he used

military force and propaganda to
claim that the rebel forces had
taken over parts of Mindanao
and the Muslim and Indigenous
regions. So he used the Army of
the Philippine Republic to fight
against Filipinx who simply wanted their homelands as their own."

Alfonso found himself involved in the armed
struggle against rebel forces in Mindanao,
though he was most likely unaware of the complete history and tangled origins of the conflict
he was participating in. Bathed and indoctrinat-
ed by nationalistic patriotism or duty to country
and family, perhaps he questioned the intent of
his service in the Philippine Army against the
Moros? Of this, we will never be sure. What is
certain are the circumstances which led to the
escalation of hostilities between the Moros and
the Philippine government. An event that paved
the way for the action Alfonso witnessed as a
young Army recruit in Mindanao.

1968 was the year that this embattled and
suppressed minority, which had struggled to

maintain autonomy and presence in the Philippine archipelago for centuries, was forced into a new phase of fighting against a bellicose entity bent on driving them out for good.

On March 18, several young Philippine Army recruits allegedly mutinied against their superiors in response to learning the objective of their recruitment and training. These Filipino-Muslim recruits were enlisted to conduct covert operations among the Moro population in preparation for the invasion of Muslim dominated Sabah by the Philippine government. Authorized by President Marcos in 1967, the intent of the then classified Operation Merdeka was to legitimize the Philippines' claim to Sabah, an area located in the northeastern portion of the island of Borneo. Malaysia, the Philippines' neighbor to the southeast, had claimed the territory as its own after officially becoming independent only four years earlier in 1963. Marcos, annoyed, believed the Philippines had a legitimate claim due to the area's inclusion in the bygone Sultanate of Sulu. His predecessor, President Diosdado Macapagal had declared his country's official stance on the matter in 1962 and Marcos, as was his custom, intended to follow up with military action.

Accounts of the events of the Jabidah massacre vary widely among many sources. Some claim as many as 68 recruits were murdered. Others document as few as 11 slayings. This confusion is because many reports and details of the incident were suppressed in the Philippine press by the secretive Marcos regime. Whatever the details of the event, what is well-known are the consequences. Majority Muslim Moros renewed their deeply embedded hatred against the Philippine government's encroachment on their perceived sovereignty with greater passion than before. The Moro National Liberation Front (MNLF) was established four years later in 1972 in response to Marcos' martial law measure and in memory of the incident. This armed group—labeled as insurgents by the Philippine government—fervently committed themselves to the establishment of an independent Muslim state in Mindanao. This led to a reinvigorated armed and cultural resistance that continues today carried out by the MNLF's many modern derivatives and the Islamic State of Iraq and Syria (ISIS).

Alfonso served in Mindanao to help suppress the rebellion of guerilla fighters on the island. With its picturesque tropical beaches and rolling mountains covered by a dense canopy of green trees, the area appears as an ideal destination for tourists to enjoy a relaxing vacation. However, Mindanao is a haven for some of the world's deadliest terrorist organizations. In the 1970s, well-armed MNLF units roamed in pocket areas of the island. For government sympathizers, naive travelers, or even other locals—especially those of the Christian faith—encounters with these forces could easily turn violent if provoked under the wrong circumstances.

The conflict intensified in 1972 after Marcos' declaration of martial law throughout the nation. The result in Mindanao was the beginning of what was historically labeled as, "The Muslim Separatist Rebellion." One source summarizes the conflict:

> From 1972 through 1976 a fero-
> cious war between Muslim sepa-

ratist rebels and the Philippine military raged throughout the southern Philippines. An estimated 120,000 people died in the fighting, which also created one million internal refugees and caused more than 100,000 Philippine Muslims to flee to Malaysia. The war was also extremely costly to the Marcos government. It was reported that, by 1975, as much as three-fourths of the Philippine Army was deployed in Muslim areas of Mindanao.

Alfonso was among the surge of military power concentrated in Mindanao during this period, though we are unsure of the exact capacity or role he played in the conflict.

The fighting officially ended in a stalemate. Marcos' military machine was unable to defeat the heavily entrenched and internationally backed Muslim rebellion. Late in 1976, the Philippine government and leaders of the MNLF gathered in Tripoli to broker a peace accord to officially end the rebellion. The result was the Tripoli Agreement. This historic agreement, "provided the general principles for Muslim autonomy in the Philippine South" as well as "...for the establishment of autonomy in the southern Philippines within the realm of sovereignty and territorial integrity of the Republic of the Philippines." The arrangement was hailed as a significant step forward between the government and Muslim citizens of the country. Despite this, fighting resumed on a much smaller scale by the end of 1977, and has continued to surface intermittently to the present day.

For young Alfonso, that was the conflict he fought in. Perhaps mirroring his own teenage years, this was a struggle for identity, sovereignty, and cultural survival centuries in the making. One that cemented military service as a prominent factor in his life for good.

———

The specifics of Alfonso's military service in the Philippines are shrouded in mystery and legend.

He rarely spoke about his experience down south to his friends or family. His sister Shirley never heard him mention details of his service, nor did his wife. We do not know the reason for this. Alfonso was quiet, even secretive by nature. Could that have contributed to him only mentioning the subject on chance occasions? How a person internalizes their military service is deeply personal. This is especially true when the veteran was involved in combat. Those intense encounters can tattoo themselves onto the psyche and, on occasion, penetrate the soul, remaining like a malignant spirit within for a long time. One wonders if Alfonso suffered from such a haunting.

Even though most of those close to him knew little to nothing about the details of his service, not everyone was completely in the dark about his life in the Philippine Army.

One of his good friends from the United States Navy, Florencio Obillo "OB" recalled in an interview that Alfonso often carried a backpack around with him when they were both young sailors. In it, he carted around all manner of supplies, food, and other utilities. When OB asked why he kept all those things with him, Alfonso told him that he always wanted to be prepared for whatever situation came his way. OB attributed the somewhat odd character quirk to Alfonso's time in the Philippine Army, where he had trained with the special forces branch of the military. "Al always wanted to be ready," OB said, with a jovial laugh. Due to his time in the Philippine special forces OB admitted he and many of Alfonso's peers respected him for having served his native country during such a tumultuous time.

After hostilities in Mindanao cooled off in the late 1970s how Alfonso spent his time between 1977 and 1979 is unknown. Most likely he returned to San Marcelino, doing humble work for a modest income. A change of pace from hectic military life probably suited him, while his family and close friends gaped at his more muscular frame and more mature disposition, unable to truly fathom the extent of all he had witnessed and experienced down south.

What we do know is that Alfonso's father possessed a green card from the United States and continued to press him to leave for America.

From Eulogio's perspective, the decision made perfect sense. By the late 1970s, the Philippines had recovered from the devastation of World War II thanks to generous government spending during the post war years. However economic, and social troubles, exacerbated by the Marcos regime's heavy-handed policies, made life tough for average Filipinos. Seeing the writing on the wall, Alfonso's father assumed his son would do better for his future to move to the U.S.A. If Alfonso had any objections to the suggestion, he did not voice them openly.

The precise date he arrived in the U.S. is unknown however, by the beginning of the next decade, Alfonso would be enlisted in the United States Navy, continuing to forge a legacy of service to a larger cause he had begun in the Philippines.

CHAPTER 3

American Sailor

*"...don't worry, your family is in good hands.
I admire you. As I've gotten older, your
friendship is one of the ones that I will
treasure for the rest
of my life."*
Manny Solano (Friend of Alfonso)

By the time Alfonso arrived at the Naval Training Center (NTC) in San Diego in July 1980 he was probably shocked by the state of the facilities.

Opened in 1923, NTC San Diego, or simply "boot camp" as it was commonly referred to, had benefited from the United States' increased need for recruits throughout the recently ended Vietnam War period. As a result of the surge

in personnel and manpower, a recently constructed eight-thousand-person mess facility
had been erected adjacent from Bainbridge
Court. This was in addition to new classes for
thirty-one apprentice class "A" and advanced
schools, administrative facilities, and most importantly, for the then 23-year-old Alfonso, new
barracks. These modern renovations had been
completed in 1970, making them still relatively
new by the time Fireman Recruit (FR) Amos lay
in his bunk, socialized, and ate there.

A photo taken inside the renovated barracks from the period displays a large bay with
two columns of single man beds in the center,
stretching back to the rear of the room. A polished gold hued vinyl floor reflects beaming flo-
rescent light from the ceiling, with additional
light and fresh air filtering in from the open
bay windows. There are a dozen beds in each
column with small shelves for linens, uniforms,
and personal items crammed in between the
two rows. A plain white sheet, pillow, an olive-
green blanket, and a set of training uniforms
are neatly folded on each bed. To curb disease
transmission, the placement of each pillow al-

ternates with the bunk adjacent to it. Alfonso would come to know this space well for several weeks in the summer of 1980, as it functioned as the hub for his military training as well as his social life.

Located on the north side of San Diego Bay and to the west of the San Diego International Airport, the training camp enjoyed ideal weather with an average daily temperature of 73 degrees Fahrenheit that slipped to about 65 degrees in the evenings. Dressed in his training uniform, a simple white T-Shirt and black pants, FR Amos probably adapted easily to the milder climate, which was much less humid than where he had spent his childhood. Despite the similarity, he probably reached for his wooly blanket at night, slightly chilly without tropical humidity hanging in the air to warm himself. The pleasant bay breeze, often churned up by the mighty Pacific and the daily passenger airliners streaking in and out overhead, made the weather one of the last things on his mind as he took on the challenging task of not only adjusting to American military life, but also growing used to the foreign culture he had suddenly been thrust into.

It is uncertain how these challenges influenced his early days in the United States, but with his prior exposure to military life in the Philippines, he likely found many aspects of his Navy training dull and repetitive. A greater challenge may have come after boot camp when he entered training to become a machinist's mate.

The rate (or "job") was established in the U.S. Navy in 1880, when it was initially known as a Finisher rate. Over twenty years later in 1904, the name machinist's mate stuck. As technology advanced throughout the twentieth and twenty-first centuries, more specialties were absorbed by the rate, but its core function remained the same: to help run and maintain "anything that moves" on a ship.

Machinist mates are vital to the functioning and maintenance of all mechanical systems onboard an operational ship. According to the Bureau of Naval Personnel (BUPERS), their job is to "operate, maintain, and repair ship propulsion machinery, auxiliary equipment, and outside machinery, such as: steering engine, hoisting machinery, food preparation equipment, refrigeration and air conditioning equipment, windlasses, elevators, and laundry equipment. Operate and maintain (organizational and Intermediate level) marine boilers, pumps, forced draft blowers, and heat exchangers; perform tests, transfers, and inventory of lubricating oils, fuels, and water. Maintain records and reports, and...perform duties in the generation and stowage of industrial gases."

Alfonso's longtime friend, Manny Solano was a fellow machinist mate. Solano, with a clear resonating voice, proudly recalled in an interview, "we are the *real* sailors...we got the toughest job." Everything with a mechanical function aboard the ship was his and Alfonso's responsibility as machinist mate.

We do not know how Alfonso initially reacted when he was introduced to the complex engineering concepts and maze-like diagrams needed to perform his duties in late 1980. Concepts including mathematics, physics, thermodynamics, mechanical theory, and fluids were included in the curriculum, forming a challenging academic undertaking for the average new sailor.

Alfonso was *not* the average sailor. He had taken trade courses as a mechanic back in the Philippines. That knowledge provided him with a solid foundation in his future technical work. Due to his mechanical experience, technical training with the Navy, or some God-given connection with machines, Alfonso grew very fond of his work as a machinist mate. Years after his passing, his son, John remarked that his dad loved to tinker with things. This interest in all machines extended to electronics and puzzles, with John fondly reflecting on the days Alfonso played Tetris with him on an old Nintendo Gameboy. A cumbersome, yet convenient device that provided hours of entertainment. Further evidence of his love of electronics exists in undated home videos. Alfonso had an epic stereo system in his home. An amplifier, tape deck, video cassette recorder (VCR) and more, form a massive black tower with knobs and lights visible from across the living room in many of the videos. It stands encased, like decorative glassware, in his entertainment center, next to his equally large television. A glowing example of a sophisticated and modern -at the time- analog sound system at the height of the home stereo era of the 1980s and 1990s.

At home and at work Alfonso demonstrated his zeal for machines. Later in his military career, he would be known for being able to fix anything on his ship, having become a skilled technician and master of his military craft.

An undated photo of him from his machinist mate days displays his comfort among the potentially chaotic, and often dangerous, lower decks of a ship. Alfonso stands centered in the frame of the shot. Surrounded by pipes of various color and thickness snaking from the polished deck to the ceiling, he poses in a relaxed manner, with his left hand gently clasped around a thin copper-toned pipe, his right resting on his thigh. A thin gold colored pipe running horizontally along the floor, acts as an improvised seat, supporting his body weight. Sporting a white t-shirt, dark blue pants, an immaculately polished pair of black boots, a black cap stitched with "USS MIDWAY" on it, and a large pair of 1980s-style glasses, he looks at home in a portion of the ship that is notorious for sweltering temperatures and cramped working spaces when underway. While the picture is undated, and we cannot be sure exactly when the photo was taken, it is possible it was snapped in 1981 or 1982, when Alfonso was assigned to the USS Midway, his first ship assignment as an active-duty sailor.

Launched on September 10, 1945, the USS Midway was a hulking aircraft carrier designed to project United States Naval power around the world. By the time Alfonso was assigned to the vessel in 1981, for nearly fifty years the float-ing fortress had been involved in several key military engagements across the planet, includ-ing the Korean and Vietnam Wars. Then later, Operation Desert storm in the 1990s. Weighing in at 65,000 tons, with a length of 1,001 feet (over 6 American football fields), and housing roughly 4,000 personnel; the Midway was an imposing force to any who saw it. Its lethal ar-mament included, 18 five-inch-fifty-four cali-ber Mark 16 guns, 84 Bofors forty-millimeter guns, Oerlikon twenty-millimeter cannons, 2 RIM-7 eight-cell Sea sparrow missile launchers, and 2 Phalanx close-in weapon system (CIWS) (*"pronounced sea-wiz"*) a newer radar guided defensive weapon with a 20-millimeter Vulcan cannon attached. The vessel relied on 12 boilers for propulsion to move its large frame across the water. These blazing hot mechanisms were vital for movement and represented a critical function in the Midway's mission across unpredictable seas. For Alfonso, they were just one of his many "projects" in the bowels of the ship.

———

Years later while assigned in San Diego to the USS Cape Cod, he met a young Florencio "OB" Billo. A man with a reserved nature, sporting clear black rimmed glasses, who is quick to let lose fits of wistful laughter, OB described Alfonso -who he referred to as "Al"- as hardworking and a tough guy. "[He] was very friendly and likeable...we had a lot of respect for him." As a fellow machinist mate and junior sailor, OB often served as Alfonso's "liberty buddy" when they were afforded the opportunity to have time away from the stresses of training and military life. This meant that the pair spent a fair amount of time socializing together. During his interview, using succinct phrases, OB recalled with laughter how a young and single Al, with a mixture of quiet charisma and his friendly smile, wooed many women. Like many military friendships, their bond was forged over late nights on the town during liberty, as well as among the pressure of high stakes military duty in which failure of any kind was not an option. The pair spent one hundred and eleven days at sea during a tense deployment in support of missions in the Middle East. An example of the time they spent together in service of their country.

Although Al was known for his calm personality, OB recounted an anecdote of a moment when Al lost his cool. Other sailors had taken one of Al's boots, leaving him looking around frantically for it. Angry, he began to shout and yell, demanding that whoever took his boot return it to him at once.

The short story bears mentioning, because it is one of the few glimpses that we have into a moment when raw emotion broke the usual waves of calm that characterized Alfonso to many of his peers and his family. OB, during his interview, reflected with a hint of sadness, "It was the first time I had seen him angry." Reminiscing on this occurrence and other events, OB seemed at ease when he mentioned the early days of his time with Al. When asked what he would say to him toward the end of his interview, OB commented with a serene smile,

> *"He's one of the best...When I see him, I [would] say to him, you did a good job."* - OB

CHAPTER 4:

A Man Defined

"I think the thing he definitely strived for was the stability with our family."
John Pamintuan Amos (Son of Alfonso)

By the time Alfonso joined the crew of the USS Midway in 1981, the ship had already been in service for nearly forty years. Much of that time was spent cruising the vast waters of the Pacific Ocean around East Asia and providing support to ongoing military operations and training exercises in the seas around Vietnam, the Korean peninsula, Japan, and many other nations in the region.

Named after the decisive Battle of Midway fought during World War II, it was constructed in only seventeen months, to speed its insertion into the American war effort at the time. Though it was unprecedented to build such a large vessel in such an abbreviated period, the Midway would miss the Earth altering war by one week, when it was commissioned on September 10, 1945. Until the end of the century, it would serve America's strategic interests abroad, leaving active service in 1992. A period that would cement its place in history as the longest-serving aircraft carrier in the twentieth century. Visitors may now freely walk the labyrinth of corridors and narrow staircases connecting its decks upon a visit to San Diego, where it sleeps, permanently moored as a museum for the public to witness a juggernaut of sea power that played an instrumental role in Naval operations throughout half of the previous century. The picture of silent cannons, cold pipes, mammoth bays cleared of operational aircraft filled to the brim with flocks of tourists hoisting smartphones, paints a stark contrast to the environment Alfonso and his shipmates inhabited when underway on the ship back in the early 1980s.

Alfonso arrived on the *Midway* just a decade after the ship had undergone an incredible four-year modernization overhaul. Troy Prince, a former Navy machinist mate, who served on the vessel in the 1990s described the details of the improvements with stunning clarity:

> "February 1966 saw *Midway* decommissioned once again in order to undergo the most extensive and complex modernization ever seen on a naval vessel. This upgrade would take four years to complete, but yielded a much more capable ship and made *Midway* operationally equivalent to the newest conventionally powered carriers. The flight deck was increased in surface area from 2.82 acres to 4.02 acres. The addition of three new deck-edge elevators could now lift 130,000 pounds compared with 74,000 pounds of her sister ships...Two powerful new catapults on the bow, three new arresting gear engines, and one barricade were installed and

rearranged to accommodate a change of 13 degrees to the angle deck...Modern electronic systems were installed, a central chilled water air conditioning system replaced hundreds of individual units, and *Midway* became the first ship to have the aviation fueling system completely converted from aviation gas to JP-5."

Alfonso found himself working on many of these modern systems, honing then eventually mastering many skills for later use during his Navy career. As a crew member of the storied vessel, he was an unknowing recipient of the supposed "Midway Magic" that endowed it with many exotic ports of call in Asia. Among these coveted destinations were Hong Kong, Perth, Singapore, Sasebo, and of course Subic Bay in his native Philippines. Between 1976 and 1983, the Midway would make a total of twenty-eight stops in the deep waters of the bay. During one of these, the mysterious wizardry attached to it would manifest in Alfonso's life in a completely unexpected way.

—

The city of Olongapo lies at the heart of Subic Bay. Today, it is a thriving metropolis with a bustling urban core that attracts thousands of vacationers annually. However, this was not always the case. Due to its location near the center of the bay, the city has long been of strategic value to the many foreign powers that have occupied the Philippines for centuries. The British, Spanish, and Americans would lay claim to the city between the mid 18th and 20th centuries. A heavy American military presence would remain after the conclusion of World War II and despite the country finally shaking free from United States imperialism in 1946, the city would continue to fall under American administrative control for another 20 years, until becoming a chartered city June 1, 1966.

When Alfonso arrived at Naval Base Subic Bay in December 1982, the port was alive with activity. Ships went out and came in. American aircraft soared over his head. Viewed from his new perspective as an American sailor and citizen, his return home must have filled him with anxious excitement.

The expansive naval station was vast and sprawling, covering 262 square miles (756 square kilometers), an area about the size of Singapore. Its size ranked it at the second largest overseas installation in the United States Armed Forces at the time, trailing only slightly behind nearby Clark Air Base 74 kilometers (about 46 miles) to the north.

It was there in Olongapo City, during a brief port call that Alfonso met Marcelina. In those days she worked as a stenographer in the city attorney's office. A pragmatic and focused young woman, Marcelina, or Marci as many of her family and friends called her, was introduced to Alfonso by one of her co-workers who also happened to be his niece. In an interview, Marci described being smitten "like she was in love" almost immediately upon meeting him. "When we first met he was telling me [that] I'm pretty. [I was] like lovestruck." Though quiet by nature, something about Marci gifted Alfonso with the courage to express himself in words that enchanted her. His flirtatious compliments combined with his radiant smile captivated her, and like magic, the two began dating shortly afterward.

Initially, they went out with other friends then, not long after, they began to spend time alone. With a slight smile, Marci recalled catching movies and bar hopping with Alfonso in the many clubs located outside of his Navy base. Though we are unsure of Alfonso's thoughts and feelings at that time, his actions indicate feelings of romance had blossomed in his heart. Somehow, Marci cracked his silent shell and earned his trust very rapidly. Marci learned he had spent time in the Philippine Army and had served in Palawan and Mindanao. She knew he had been involved in the conflict against the insurgents and was well trained in how to kill people. Though she felt if he had taken any lives, he had done so to protect himself.

Alfonso and Marci's romance that Spring was intense but short-lived, as Alfonso would leave Olongapo City only ten days after they began dating. The *Midway* was due to leave port and duty called. Despite the fleeting courtship, the two had formed an inseparable bond anchored in deep feelings of trust and respect. This connection only intensified over the next seven months as they continued to correspond via love letters.

Marci, her wireframe glasses set low on her nose, wearing her red t-shirt with the face of Minnie Mouse printed on it, recounted in how modern communication technology like email didn't exist in 1982, so written letters through slow going mail was the only way for her to stay in touch with Alfonso. When he returned to the Philippines that December, the two were married in a civil ceremony at Marci's very own office in Olongapo City. They held their official wedding ceremony just two months later, on February 20, 1983.

Photos from the celebration tell us that the newlywed couple had a multitude of family, friends, and likely other community members, who attended to wish them well. Marci looks stunning in her simple, yet elegant white wedding dress. Alfonso stands next to her in a collared white dress shirt and black slacks. In one picture, both wear stoic expressions as they stand among bridesmaids and groomsmen for a picture of the wedding party, as onlookers view them from the church pews. For Marci, her marriage to Alfonso was the beginning of a brand-new life of adventure. She would move to the United States a year later to follow Alfonso to his second tour of duty, cementing her status as a Navy military spouse. A role that, unknown to her at the time, was a job in and of itself. A job filled with challenges and hardships she would soon come to know and understand intimately well.

———

After months traveling the seas with the USS Midway Alfonso was reassigned to the USS Enterprise CVN-65, in 1982. The world's first nuclear powered aircraft carrier, it was commissioned in November 1961. At the time of Alfonso's arrival, the homeport of the "Big E," the ship's nickname, was in Alameda, California. Located in the heart of the San Francisco

Bay area, east of the mighty city of San Francisco, the town is comprised of three small bay islands. Alfonso would come to know it and the area around it very well in the coming years.

Marci joined her husband there a few months after they were married. Once situated in their new location, the young couple began their military life. Alfonso continued to deploy with his ship on a regular basis, while the ever-industrious Marci worked two jobs to, in her words, "make myself busy," in nearby Oakland while furthering her education. With both of their schedules full, their home life was typical of a military couple, with both passing like ships in the night on most days of the week, while stealing precious moments together whenever Alfonso had a lull in his duty schedule and Marci was free from work and school.

Things became more complex when Marci discovered she was pregnant. The details of the pregnancy as well as Alfonso's feelings on the matter have been lost to history, but Marci described in unadorned language how she lost the baby because she missed a critical wellness appointment. One can surmise that both of their frenetic schedules contributed to the oversight, which for a couple under normal circumstances would have been a pressing priority.

Despite the tragic event, the family would remain resilient and eventually celebrate the arrival of their first child, John "JP" Amos, on May 11, 1986. Nearly a year and a half later, the couple's first daughter, Jennifer Pamintuan Amos, would join the family on December 15, 1987. Their third and final child, Josephine Pamintuan Amos, would cap off the trio of siblings, arriving on February 28, 1993. The three children would occupy large spaces in Alfonso and Marci's hearts from the moment they drew their first breaths. All would be instrumental in passing on the legacy of their father by contributing to this biography.

As for Alfonso, one wonders how the loss of his first child affected his emotions. Marci hinted that despite his initial comfort and openness around her, over the years she would struggle to know her husband's innermost thoughts or feelings. Many of Alfonso's friends and family expressed similar experiences. Most described him as a quiet person who smiled a lot, remembering an agreeable sailor, father, and friend who mostly rode the highs and lows of life with relative ease. At the time, few could fathom potential sources of discomfort or pain that might be stirring behind his friendly smile, or ways that they might sweep away the effects of accumulating storm clouds over his psyche.

———

While Alfonso lived in California, he still managed to find fleeting moments for leisure with his family and for himself. An acquaintance, Wyne Corpus, who also resided in the Bay Area at the time in the early 1980s, described how he would pick Alfonso up from Alameda Naval Air Station and spend time with him occasionally on weekends. He recalled in an email, "Often times he requested Filipino food during his visit and most of the time he asked [my] mom to cook mongo beans with ampalaya and pork." Corpus also described how on Memorial Day weekend in 1983, Alfonso visited him in Vallejo, a small town about a 45-minute drive to the north. Alfonso and his brother-in-law went to find a goat to purchase in Stockton, situated about 75 miles (121 kilometers) east of the Alameda. The story is incomplete, but it is revealing that Alfonso would travel such a distance to find an animal for a get together.

Alfonso had a strong affinity with animals. From an early age, he helped his father maintain their farmland back in the Philippines. Corpus' family lived in the same plaza as Alfonso's in Linusungan, a barangay or village, located in their home country. He remembered a young Alfonso, riding astride a carabao -a swamp-type water buffalo native to the Philippines- to check on the farms in the morning, only to return on the beast hours later by lunchtime. Alfonso's fondness for animals would extend into his adult life. His eldest daughter, Jennifer, recounted early memories of her father slaughtering animals, saying it was normal for him and her mom to "raise what they eat." She elaborated:

> "I just remember when we used to live in San Diego, we had a whole like [sic] chicken coop in our backyard and it was just so nor-

mal for my dad to just kill animals in front of me with like [sic] no remorse...he would cut the head off of a chicken...and it'd be literally running around headless... that was normal for him."

His son, John, similarly recalled his father's tendency to keep animals around the house, mentioning he "built the little hen house in our backyard...we had like [sic] little chicks and hens in the backyard and I think we had several pets." John also hinted his father may have collected animals so that he could use them in meals. While we have no explicit evidence of this, information from another friend of Alfonso's, Melchor Quitoriano, seems to corroborate the claim.

Quitoriano met Alfonso when the two were stationed on the *USS Midway* together. During his interview, Quitoriano described how Alfonso occasionally became an impromptu chef for him and others when they lived in the dorms, endearing him to the rest of his fellow junior sailors. He mentioned how Alfonso sometimes liked to cook improvised Filipino dishes in a rice cooker when they both lived in the barracks. One of those dishes was papaitan, a traditional Ilocano delicacy typically made using the innards from a goat.

On a visit to the Philippines with Alfonso, Quitoriano described a particularly wild night where the entire village erupted in a festive celebration which began with the slaying of a goat. He recalled with a mischievous snicker how Alfonso took delight in helping him kill the animal. Whether or not Alfonso slaughtered the goat for tradition or to provide ingredients for food, we will never know. However, it is clear his upbringing as the son of a farmer, for better or worse, influenced his relationship with animals. A bond to his homeland that through his cooking, fishing, and raising of animals remained a feature of his personality for the rest of his life.

CHAPTER 5

Peaceful Days

"When we went to Mount Fuji and
Toshimaen, those are the times that
[I remember.]"
Lucila (Lito) Solano (Friend of Alfonso)

Alfonso relocated to Yokohama, Japan in late 1984. He had been stationed there during his first tour following bootcamp between November 1980 and 1982, so he was already familiar with life in the country.

Quietly tucked away in Japan's second largest city, the U.S. Naval Housing Annex Negishi was a stark contrast from the hustle and action of the Tokyo metropolitan area surrounding it. The houses are spread out, imitating American suburbia with large plots of land reserved for each house. Grassy lawns and trees that sprout cherry blossoms every Spring populate the area, offsetting Tokyo's grey urban sprawl outside of the gate.

It was here that Alfonso began a 36-month shore duty rotation that was likely a welcome change of pace from the heavy operations tempo he had experienced in California. Following this tour, he was reassigned to Commander Fleet Activities Yokosuka and would spend another two years there, only to return again in 1995, where he served until his disappearance. Of his eighteen years in service, he spent half of them stationed in The Land of the Rising Sun. He came to know the area very well.

While we don't have Alfonso's thoughts about living and working in Japan, his actions suggest that he felt at home there. Marci commented her husband kept cats and birds around the house, and planted things in the yard. When Alfonso was comfortable enough, he fell back on his farming roots, raising animals, planting, and watching things grow. That nurturing instinct extended to his family. Yokohama was the place where his first two children, John and Jennifer, were born.

Fatherhood appears to have been a natural fit for Alfonso. Francisco (Efren) Pamintuan, his brother-in-law, mentioned how great of a dad Alfonso was to his children. He remarked with a gleam in his eyes, "He played with his kids a lot." Efren recalled how Alfonso used terms of endearment in Ilocano to talk to them and when they grew older, encouraged them to go to college and have a good life. Even though he wasn't known to talk very much, he expressed care through his presence when he could. An admirable trait most of his family and friends mentioned during their interviews.

Home videos of him with his children provide visual documentation of the steadfast influence he had on their lives. In one, young JP and Jennifer are playing in a pool. While they whoop and splash water every which way, Alfonso roves like a sentry around the pool, watching them to ensure their safety. He then wordlessly scoops JP's skinny frame out of the pool, only to place him back in seconds later after he wriggles and worms around in his grip. JP happily returns to swimming and splashing about. There are other examples of Alfonso doing dad things, but this one illustrates how his presence as a father was less about what he said, and more about the actions he took to be present for his children and family.

———

During his second tour in Japan in the early 1990s, Alfonso reunited with his old buddy, Manny Solano. They had initially met when they were stationed on the USS Cape Cod in San Diego. He and Alfonso were good friends who enjoyed many fun off-duty activities together. On a brief port call to Dubai, they went four-wheeling among the desert dunes and rode camels. They also fished, drank, and ate together in their spare time. The memories cascaded into one another as Manny described them with excited bursts of speech.

The pair must have complemented each other well. Manny, with his outgoing and boisterous nature, was a balancing force for the quiet, soft-spoken Alfonso. Their time together was probably deeply powerful and beneficial for both men and their families.

Al and Manny took part-time jobs as taxi drivers around Yokosuka Naval Base to make some extra money. One evening, Manny waited in his car along with a long line of other drivers waiting for fares outside of the base main gate. Suddenly, a muffled voice sounded from his taxi radio, *"Number 46, sound off!"* No response came from any of the other drivers.

"Number 46, where are you!?" The dispatcher repeated, irritation rising in his voice. Manny and the other drivers were bewildered at the unusual lack of response. Seconds later, he realized 46 was *Al's* taxi number. *Did something happen to him? Where is he?* Manny wondered.

The dispatcher's voice blared from the radio, *"All taxi drivers be advised. Be on the lookout for number 46."*

Soon Manny and the other taxis were searching for him. After a few minutes Manny found his taxi parked behind the Bachelor Enlisted Quarters (BEQ). He went up to his driver side window and peered in. There he saw Al, in his words, "sleeping like a baby." Manny knocked on the window.

"Al! The dispatcher's been calling you!"

He jolted awake and immediately returned to his taxi duty. Among the other drivers, he became known as "Sleepy Al" from that day forward.

Retelling the story, Manny burst out laughing. "I think he got suspended for like one week for doing that!"

Alfonso, Manny, and other friends they were stationed with in Japan took full advantage of the free attraction tickets gifted to them by the Navy. They took regular trips to Tokyo Disneyland, Toshimaen Amusement Park, and Tokyo Summerland together with their families.

Manny's wife, Lucila Solano, recalled those days with wistful eyes. She had grown up in Zambales back in the Philippines, just like Alfonso, although the two didn't meet until he was stationed in San Diego in the early 1990s. Lucila remembered how she and Al got along well, because they were both "happy-go-lucky" people. In Japan, both families had small children, so it was common for them to take free buses to various tourist destinations around the base and enjoy themselves. She described those peaceful days:

> "We were always together, you know? Gatherings especially with MWR, [sic] have this special trip for the military families, so your family, [speaking to Alfonso's youngest daughter, Josephine] the Ava's family, and our family usually ride the bus, the free bus going to wherever...so we had fun."

Lucila wished to pass the following message to Alfonso and Marci.

> "I know you sacrifice a lot and bringing the children to the States [and] that you made it. You made it [and] you're successful. And I'm sure if Al is out there in Heaven...I'm sure he is happy that you did your best Marci. I know

you're a strong woman and I just
hope someday, you'll find happiness too." - Lucila

Her words rang with sincerity and compas-
sion for Marci and her lost husband.

―――

From all appearances, Alfonso's life was in a
period of plenty during his first tour in Japan.
Such was the nature of shore duty assignments.
The general slowness of the operations tempo
afforded him time. He made friends, was able
to be present for his family and was able to visit
the Philippines more often.

Traveling Space-Available was another one
of the many perks he and Marci took advantage
of while being stationed in Japan. The program
allows active-duty personnel and dependents to
travel on military aircraft, given that there is
space available on certain flights after mission
requirements for transporting gear, supplies,
and mission critical personnel are met. These
free flights rotate regularly between Ameri-

can bases in the region. Back in the late 1980s, flights to Clark Air Base from the busy Pacific hub at Yokota Air Base near Tokyo, were frequent. This allowed Alfonso and his family to visit his home country every year during that first tour.

Several of these trips were recorded via an old camcorder. In one, Alfonso wades in waste deep water while baby JP sits in a brightly colored inflatable raft. Father and son both smile as Marci watches from the nearby shore. These periods were probably special for Alfonso. He was able to take much needed time away from Navy life, reconnect with friends and family back in the Philippines, and introduce his children to the culture and environment he was raised in. As few as these videos may be, they provide a window into his quiet soul.

Another video clip captured Alfonso speaking on camera. It was recorded in 1986, at the baptism ceremony for his son John. He stands in a neatly pressed collared white shirt and khaki pants. Marci is by his side dressed in white as well, holding a plump baby John in her arms. The two are joined by other friends and family members in a small room in a chapel as they make small talk with the priest regarding the origin of baby John's name. His hands in a fig-leaf position, eyes shifting from side-to-side, a shaky smile crosses Alfonso's face when he says,

> "Last time I wanted just Paul, then we add [sic] John, so they came up together."

The priest responds. "I wasn't sure if you named him after John Paul because he's gonna be the next Pope. Or your Navy background after John Paul Jones." Alfonso laughed at the joke.

The rare moment captured on camera is notable for two reasons. One, it offers an uncommon glimpse into Alfonso's subdued persona. Two, it is visual proof he preferred to remain in the background and being the focus of attention made him uneasy and anxious. One wonders if these factors might have contributed to underlying depression, gnawing at him from within, where no one else could see or provide reassurance. On that point, we'll never be sure.

———

Those peaceful days in Japan must have felt like pure bliss for Alfonso and his family. But by November 1987, his 36 months had come and gone. It was time for him to rotate back to sea duty, which meant reassignment to a new post. Or so he thought. In a twist of fate, he was surprised to learn that his new post would be in Yokosuka, Japan, only 33 kilometers south of Yokohama. What was more shocking was that he was once again assigned to the USS Midway.

Even though he was probably very familiar with his previous duties on the vessel, he would still have to complete training before he returned to life at sea. After the new year, on January 4, 1988, he returned to San Diego to complete Leadership and Management Education Training. Now a Petty Officer Second Class, he was moving up in rank and responsibility, and the course helped expand his skills as a seasoned non-commissioned officer (NCO) and sailor.

We have no record of how the course went for Alfonso or what other stressors were going on in his life at the time, but we know he completed it on January 15, 1988. Two weeks later, he returned to Japan, then immediately shipped out to sea for the first time in several years.

This time was different than before. Something was stirring in his mind that didn't sit right with him. A combination of factors that were churning as violently as the ocean during the formation of a typhoon. Forces that were positioned to break his usual smiling exterior and unleash a tidal wave of emotion he was not prepared to endure.

Civil wedding 1982 Olongapo City

*Wedding in St. Joseph Church in Olongapo City
February 20, 1983 (left to right) Auntie Shirley
sister to Alfonso, unknown godmother of the
wedding (wedding sponsor), Alfonso, Marci,
Uncle Efren brother to Marci, LeiLei daughter
of Auntie Del Marci's sister, Venalynn daughter
of Uncle Ven Marci's brother, Oliver son of Uncle
Ven Marci's brother, Alan son of Auntie Del*

Alfonso's life as a Machinist Mate kept him working on the mechanical systems of the ships he was assigned to. Hot, dirty, stressful work.

Alfonso found respite in music, whether playing and singing or listening

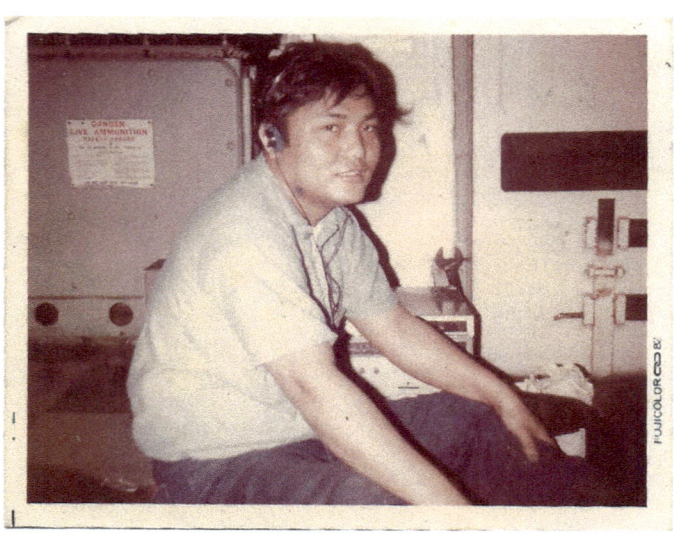

*The primary way to stay in touch when
deployed in the 1980's – by payphone*

Alfonso on the Flight Deck of one of the several aircraft carriers he was assigned to.

Alfonso at the Navy's Air Conditioning and Refrigeration School, Aug 1983

*Alfonso at the Navy's Air Conditioning and
Refrigeration School, Aug-Oct 1983*

Alfonso and Marci Oakland, CA 1984

Marci and Alfonso with friends at Christmas,
Oakland, California 1983/84

*Marci and Alfonso visiting Sea
World San Diego, CA 1984*

*Mount Fuji Camp Ground, Japan 1988
Alfonso's shipmates and their kids, Uncle
Melchor, John, Jen, Marci, Alfonso*

*Camp John Hay, Baguio City 1989 Uncle Efren
Marci's brother, Alfonso, Marci, John, Jen, Grandma
Pilar, Grandpa Segundo Marci's dad and his
friends, Uncle Ven Marci's brother and his family*

*Marci, Jen, Alfonso, John Baguio
City, Philippines 1989*

Family Portrait San Diego, CA 1992
Marci, Alfonso, John, Jen

*Marci, Josephine, Alfonso, and Jen at
Imperial Beach, San Diego, CA 1993*

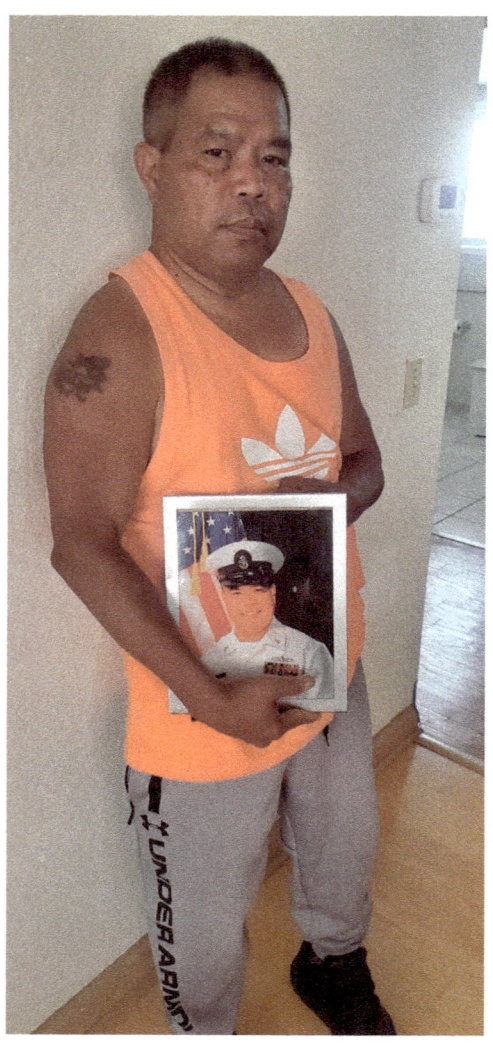

*Melchor Quitoriano, Alfonso's best friend,
with his final official Navy photo. Photo taken
during interview for this book in July 2020.*

CHAPTER 6

Going Through Darkness

---

*"I think that he was going through so much
and not being near family that maybe,
he wanted to end it all."*
Jennifer Pamintuan Amos
(Alfonso's Eldest Daughter)

The return to sea was abrupt. For Alfonso, it probably made the last few years of bliss in Japan feel like a foggy dream. Now, he had returned to the waking world. A cramped and burning existence in the bowels of a ship, with 12 hours shifts, watch duties, and stomach-turning food. It was his own personal hell. Worse still was that his young family was thousands of miles away. There were, of course, sporadic letters he relied upon to stay in touch, but it was no substitute for being there for them. He was alone and stressed, with no idea how to deal with anxiety.

Mental health and suicide awareness are common topics in all branches of the U.S. military in 2022. But that wasn't the case back in 1988. Cultural standards of the day frowned upon men who showed too much emotion or openly voiced unresolved personal issues on a regular basis. There *were* active suicide prevention programs at the time, but they were not as utilized by sailors compared to more contemporary times. Furthermore, many military personnel avoided turning to official channels for help with psychological issues. This was due to fear of jeopardizing their career, or fear of being ostracized by co-workers and supervisors who might begin to treat them differently because of their problems.

Therefore, many concerns related to mental health tended to be handled quietly through unofficial channels at the lowest level possible. Many were never addressed because the sailors who suffered in silence with them dared not uttered a word to anyone about it. They kept their feelings locked up tight away from command, friends, family, and even themselves. This is what happened to Alfonso.

Suicide is a complex topic. When a person's mind is assaulted on all fronts by a combination of factors, it can arise as an option to free them or those closest to them, from seemingly insurmountable burdens or worries.

The exact motives why Alfonso attempted to end his life all those years ago are unknown to us. He left no writings or records for us to analyze or have a glimpse into his psyche. Even with such documentation, it would be difficult to know or understand the true cause behind the act, as many suicides are often sparked by a spontaneous decision to escape lingering pain, guilt, or some other unresolved emotion.

In this work, there is only informed conjecture of his motivation. A collection of stories, anecdotes, and signs that possibly drove him to commit a dire act with permanent consequences.

———

Family issues are often a source of stress for military members. Extended time away from home and the ever-present nature of the mission, can make even the smallest problems balloon into significant issues by the time both partners have the bandwidth and time to address them. This is especially true if open and regular communication is difficult for one or both parties.

Like all couples, Alfonso and Marci had aspects of their marriage that were difficult to navigate. These choppy patches represented challenges on their course through military life.

In their younger days, Marci was a spender, Alfonso was a saver. This small stressor may have contributed to his psychological burden.

Then there was the issue of Marci's citizenship. Three years after arriving in the United States the question of when she should apply was a sore subject for Alfonso, who wanted her to wait for reasons that are unknown to us. Marci was a planner who didn't like to leave things to chance. She was a young immigrant who had abandoned her life in the Philippines to move across the Pacific Ocean with her U.S. Navy husband. She was in an extremely vulnerable position should the worst happen between her and Alfonso. Stability was her goal and getting her U.S. citizenship was the fastest way to achieve that. Eventually she did receive her green card, but the entire issue was probably another family stressor that Alfonso silently ruminated on at the time.

Years later in the fall of 1993, while living in San Diego, Marci wanted Alfonso to bring her mother stateside from the Philippines. Alfonso refused, because he felt it wasn't financially feasible at the time. This upset Marci, who was already fuming due to a recent argument with Alfonso's mother, Dolores. She had forgotten to pick their kids up from school one day and Marci was livid because of it. The altercation escalated enough to where Dolores left their house to go stay with Alfonso's brother, Rhoger, who lived in the area.

With his wife and mother at odds with each other, Alfonso began to feel unwell physically. He expressed he felt excessively stressed, was sleeping poorly, and was suffering from frequent headaches among other nagging medical concerns. He even requested medication to help calm his nerves, an uncharacteristic action for him. A doctor diagnosed him with depression and suggested marriage counseling. It is not clear if he ever attended.

His family was always at the front of his mind, so the trouble at home was a significant source of stress for Alfonso.

Health issues may have also contributed to Alfonso's depression. He smoked regularly until 1996 when he quit suddenly at the age of 42. In 1987, he was seen three times throughout the year for flu like symptoms. He had never been sick that often before. Although the smoking was probably not the sole cause of his recurring illnesses, it could have been a contributing factor, and as a result, could have become a stressor he worried about.

His shore and ship rotation may have been factors that influenced his health as well. Jennifer, his eldest daughter, remembered how his appearance was noticeably different when he had been away and when he was at home for a longer period of time. Sometimes he was lean and somewhat fit following deployments. Other times his physique would soften and become round after spending time at home. While the weight fluctuations and the medical issues do not say much in isolation, could they have affected Alfonso's view of himself and his abilities to perform his job?

———

Alfonso's job as a machinist mate was stressful. As a young sailor he was responsible for knowing the inner anatomy of the ship inside and out in order to ensure all heating, cooling, and other mechanical systems hummed along so the ship would be always mission capable.

His brother-in-law, Efren, acknowledged the stress of the job describing it as "hot, greasy, dirty" work. Others echoed the demands of the rate in their interviews, with the *heat* being the recurring feature.

The hot temperatures of his workspace, combined with the high pressure to perform in a new role, may have overloaded his resistance to stress. For two weeks in early January 1988, he attended a leadership course. What he learned at the course is unknown, but given his time in service up to that point (roughly 8 years), the training was probably meant to prepare him for a position that was more supervisory in nature. He was no longer a junior sailor only responsible for the maintenance of the moving parts of the ship. He would now be leading other sailors at sea, a responsibility many NCOs accept with anxious eagerness. Did Alfonso shoulder the job with that same sentiment?

———

When Petty Officer Second Class Amos returned to Japan toward the end of that month, he likely had little time at home before going underway. Trouble was brewing in the Middle East between Iraq and Iran. His ship would need to be on standby in the Arabian Sea within a few days in order to assist what had been dubbed Operation EARNEST WILL. The primary objective was to protect Kuwaiti owned oil tankers from Iranian attacks in the Persian Gulf. This was no exercise. It was the real deal and Alfonso was right in the center of it.

This time, getting the ship underway was probably more nerve-wracking and demanding than ever before. The circumstances must have sparked unimaginable tension for Alfonso. It was his first time going to sea in over three years. He had a new position, likely a new crew, and a live operation to prepare for, in addition to recalling and utilizing all of his previous machinist mate training to ensure all systems were good to go so the ship could get going as soon as possible.

Cameron McCulloch, a Mass Communication Specialist 3rd Class, wrote about the arduous processes those in the engineering department must go through in order to accomplish their mission. "...To go underway, engineering department Sailors are working long hours, deep in the bowels of the ship, ensuring the plants, generators and evaporators are ready when it's time to deploy...all that power giving equipment requires regular maintenance to keep the ship operational and mission ready." There was little rest for Alfonso during this period. Operational tests needed to be performed. Last minute issues with faulty systems fixed. The entire movement of the ship rested on the blackened hands of his and others in the engineering department. After a grueling few days of non-stop work everything was ready. Due to Alfonso and his shipmates' efforts the Midway sailed away from Japan, reaching its destination in the North Arabian Sea several days later.

—

Early in the morning on January 29, 1988, Alfonso lay in his rack unable to sleep. Jumbled thoughts swirled in his mind. Marci, little JP, and baby Jennifer. *How were they?* His health. *Will I be ok to do my job?* The job. *Can I do this?* The thoughts drew tracks circling in his head. Then, without much thought, he took a new razor blade and ran it across his right wrist.

As the warm blood began to flow out and onto his sheets, one can only imagine the void he was in. A black mass he hoped would swallow him and his worries whole.

He lay there floating, alone in the noiseless sea of his own misery for four hours.

Then suddenly, he decided he wanted to live. He sat up, gingerly wrapped his wrist, then reported to work. Two and half hours later, a fellow shipmate noticed the bloodstained sheets of his rack and reported it to his supervisor. That was when Alfonso was rushed to the ship doctor to immediately patch the wound.

———

What followed for Alfonso was a whirlwind of visitors, evaluations, movements, and procedures. After his wrist had been sufficiently stabilized, he was airlifted back to Yokosuka the next day and referred to psychiatry for a fitness for duty assessment. When asked why he attempted to end his life by various medical authorities, he cited being worried about his family and concern that he would not be able to perform his new job adequately as the primary reasons. Soon after his evals were complete and his wrist had healed enough, he was declared fit for duty again and returned to his Navy life.

———

Alfonso never spoke of his suicide attempt to anyone in much detail. After being discharged from the Yokosuka Naval Hospital, he returned home with a cast. Marci knew he was stressed about starting a new job and leaving baby Jennifer, who was 2-months-old at the time, but had no idea it was bad enough for him to hurt himself. She said it looked like he was carrying a heavy burden on his shoulders. One she did not know how to relieve him of.

Josephine Amos, Alfonso's youngest daughter, reflected on the incident.

> "He didn't explain himself [to his brother or anyone]. And then he ended up going back out on another ship. A couple days after?
>
> ...Finding that out really, really hurt. Just imagining the amount of pain he must have been in and feeling like he had no other options. Or maybe he was battling depression. It just didn't seem like he opened up to anyone."

During the same conversation, Efren, Alfonso's brother-in-law who previously had no knowledge of the suicide attempt before the interview, added:

> "As I mentioned, he kept to himself. He's soft spoken. You know... most men are like that. Because there's that tendency to feel that if you bare your heart up, that's a weakness. During that time, none of that. No counseling. You don't go to a psychologist or psychiatrist. Because you're afraid that your career is going to get affected. So you keep it to yourself and to your best friend. Sometimes men even hide it from their wives. Because that fear of you don't want to be labeled as you're weak. You know? You know when people experience what they call dark night of the soul. And everybody has to go through that one way or another, depending on how extreme, the

severity of it. But yeah, we could only imagine when people...go through darkness."

Even though the Navy had deemed him fit for duty, something about that darkness never left Alfonso after that point. Perhaps it had always been there festering like a cancer, only becoming visible after a magnificent load of stress taxed his body and mind sufficiently. No one knows. But in many ways, he wasn't done with the darkness. And it wasn't done with him.

CHAPTER 7

Career of a Machinist Mate

---

*"We were all machinist mates and [he] was
an expert at finding parts. And he's good at
the computer. No Google whatsoever.
The microfiche was his expertise."*
Manny Solano
(Alfonso's Friend, fellow Machinist Mate)

Alfonso's relationship with his work as a machinist mate is difficult to understand. Accounts from loved ones suggest that he was excellent on the job, taking advantage of innate talent and interest in the technology of his day to excel in his respective roles.

However, being skilled at something is quite different from enjoying it. Like many other areas of his life, work was a topic he rarely discussed with anyone. Especially those who were outside of his professional sphere. Few who knew him could speak with knowledgeable accuracy on his feelings about his job in the Navy. They were aware he worked hard all the time, but could not elaborate any further.

The nature of his work could be physically and mentally taxing. Cramped spaces, blazing hot temperatures, roaring turbines, decks slick with grease, and other environmental hazards, made every day at sea treacherous. Injuries were common for Alfonso and others who labored alongside him in such conditions.

Early in his career, dust particles scratched his eye, forcing him to seek medical attention. A few years later, a ladder he was walking down fell over and badly bruised his leg. These were just two recorded work mishaps. He undoubtedly sustained other less severe injuries while repairing equipment or traversing under inconveniently low ship passageways between dimly lit corridors. The risk of bumping his head against solid steel was always present and just one of the occupational risks of ship life. He encountered this hazard and more every time he went to sea.

—

Fireman Apprentice (FA) Amos departed to sea on his first ship deployment March 12, 1981. The ship to which he was assigned at the time, the USS Midway, departed Yokosuka and sailed for the Indian Ocean. Excited and somewhat nervous, he embarked on his first adventure as a United States Navy Sailor with little clue of how exactly the deployment would play out.

Two days prior to the ship's departure, a civilian helicopter had gone down over the South China Sea. The helicopter had been ferrying seventeen passengers from Indonesia, Japan, Singapore, The Netherlands, and the United States. For days, the wreckage and the passengers could not be found among the vast 1,423,000 square miles (3,685,000 square kilometers) of the vast sea. All seemed lost until the Midway, with Alfonso aboard arrived at the scene.

Troy Prince, a former Machinist Mate and USS Midway historian writes:

"On March 16, 1981, seventeen passengers of a crashed civilian helicopter were rescued by men of the USS *Midway*. The survivors (Indonesian, Japanese, Singaporeans, Dutch, and American) were first discovered by an A-6 Intruder from VA-115 during a routine surface search mission in the South China Sea. *Midway* immediately dispatched helicopters from HC-1 Det 2 to the scene and all seventeen people aboard the downed helicopter were rescued and brought aboard the carrier where they received medical treatment and food. The chartered civilian helicopter was also plucked out of the water and lifted to *Midway's* flight deck. Two days later, the ship entered port at Singapore, home for most of the survivors."

After dropping the passengers off, the Midway continued to its destination in the Indian Ocean, then returned to Japan 5 weeks later. Alfonso was awarded his first Navy Expeditionary Medal and Sea Service Deployment Ribbon for his contributions while at sea.

———

His second and third deployments were aboard the USS Enterprise, when Alfonso was based in Alameda, California at the small Navy installation there. With more training and experience under his belt, Fireman Amos felt much more prepared to go to sea than before.

The first of the two was back to the Indian Ocean to participate in WESTPAC, a large deployment exercise. Alfonso departed in September 1982. He visited many coveted ports of call during this deployment. Hawaii, Singapore, Subic Bay (where he met his future wife Marci), Mombasa in Kenya, Australia, the Philippines a second time in early 1983, and Sasebo in Japan, before returning home. The trip was eye opening for him as he took in parts of the world he had only heard or read about in the past.

After a lengthy eight-months, his Pacific tour was complete and Alfonso headed home on the Enterprise. All was well until the ship arrived back in U.S. waters. Disaster struck when the hulking 93,284-ton aircraft carrier ran aground after crossing under the Golden Gate Bridge in San Francisco Bay, a mere 3,000 feet (914 meters) from the pier.

George C. Wilson, a reporter at the time, wrote about the potential cause of the embarrassing blunder in The Washington Post:

> "The helmsman apparently lost control of the ship after the farthest out of its two starboard propellers stopped churning, causing the two port-side propellers to exert more lateral force than the rudder could overcome before the giant ship went aground, Navy officials said... An official of the San Francisco Bar Pilots Association said a civilian pilot had been at

the carrier's helm as it sailed under the Golden Gate but he turned over the wheel to a Navy pilot just before the ship went aground, according to the Associated Press... At one point during the Enterprise's grounding, witnesses said, crew members were ordered to gather in one part of the ship, in hopes that their weight would tilt the carrier enough to make it slide off the mud or would correct the listing. Both unofficial explanations were offered at the scene."

The humorous incident was no laughing matter for Captain R.J. Kelley, but for Alfonso and the rest of the crew, it became a popular sea story for the rest of their careers.

———

A few months after returning from the deployment, Alfonso was promoted to the rank of Machinist Mate 3rd Class (MM3). As in all branches of the U.S. military, his new position as an NCO gave him more responsibility and a noticeable bump in pay. The latter was more important in Alfonso's eyes.

A year later, on May 30, 1984, Machinist Mate 3rd Class Amos departed on his second and final deployment on the Enterprise. The enviable itinerary saw him visit some of the most coveted ports in the world.

Alfonso enjoyed port calls all throughout the Pacific. He participated in RIMPAC 84' at Pearl Harbor, Hawaii on a 4-day port call. From there, it was a 9-day visit to Naval Air Station Cubi Point in his home country the Philippines. He left for Hong Kong and arrived there in early August for a 5-day stay. A few weeks later, the Enterprise arrived in the North Arabian Sea, where it remained for 2 months. Next, it was back to Subic Bay in the Philippines for a week, then Pearl Harbor one more time for 3 days before returning home to California.

These two deployments were treasured experiences for Alfonso. He got to see more places around the world than the average person sees in a lifetime. He met his wife. And he gained valuable work experience participating in the various exercises and battle drills at sea.

Yet, while these episodes would have been viewed as fabulously positive by the average sailor, for Alfonso he had several things on his mind throughout the duration of both deployments. He was still new to living among American culture and had difficulty integrating into his unit which was filled with American-born, primarily white sailors. Manny, his loyal friend and fellow machinist mate, recalled being made fun of because of his accent. Efren, Alfonso's brother-in-law, was denied leave by a white officer because he was Filipino. When he went above the officer to his supervisor and his leave was approved; the officer became furious and yelled at him in a secluded corner of his ship.

Both Manny and Efren implied that if they had been discriminated against during their time in service, it was likely Alfonso had been too. Though if he did encounter explicit prejudice, he rarely discussed it with either of them.

During shore duty assignments Alfonso focused less on work and more on his family. These joyous periods afforded him time to do so. Because his work became subordinate to his family life during these times, there is little documentation of his work performance at these posts. Nor is there any documented record of his opinions or feelings about his job. He did what he had to do, then went home during those times, no more no less. He did continue to keep up with mandatory training whenever they sent him, but the longer courses tended to be in-between assignments. There's no indication he participated in any additional training outside of them.

In July 1991, he was assigned to the USS Cape Cod in San Diego. A Yellowstone class destroyer depot ship, the Cape Cod was a smaller vessel than Alfonso was used to, which meant his job was less demanding than when he worked on the Midway or Enterprise.

The USS Cape Cod was an auxiliary class ship that saw less use as the 20th century neared its end. There is a record of the ship deploying to the Philippines in June 1991 in support of the cataclysmic eruption of Mount Pinatubo. However, Alfonso would not arrive until a month later, missing out on the opportunity to be involved with the historic event and return home to ensure the safety of family and friends who lived less than 100 kilometers (62 miles) away.

As the years rolled on in San Diego, Alfonso seemed content with his life there. In several home videos recorded during that time, he is on camera shooting pool, eating, and even dancing. Relaxed in a white buttoned shirt adorned with tropical flowers, he breaks out a spontaneous dance move as The Ad Libs *Boy From New York City* plays on the speakers wearing an unmistakable look of joy on his face.

His youngest daughter was born two years later. February 28, 1993, Josephine Pamintuan Amos joined the family. Shortly after her birth, Marci captured Alfonso on video as he held his new daughter. His face was expressionless. The eyes bleary as he looked toward a point off camera. He appeared exhausted. Whether it was from a long night of sitting by his wife's side as she gave birth or for other reasons wasn't clear. But it was a look he wore often in that and many other family videos.

———

Before his idyllic assignment in San Diego ended, Alfonso completed a two-week certification course for processing refrigerants. Training meant to increase his capabilities as a machinist mate. Up to that point, he hadn't completed any documented training for his job since his arrival at the assignment in 1991. In his mind, the message from the Navy was clear: ready or not, prepare for sea duty again.

By October 1995, he was back at Yokosuka, Japan, but this time he was assigned to the legendary USS Independence. Nearing the end of its service life, the storied aircraft carrier participated in many of the most historic and significant U.S. naval engagements throughout the twentieth century. The ship participated in the Cuban Missile Crisis, Vietnam War, operations in Grenada, Desert Shield, Operation Southern Watch, countless exercises, operations in the frosty Arctic then Antarctic circles and everywhere in between. It projected U.S. Navy seapower around the globe and was an active participant in determining the outcome of pivotal military actions throughout the second half of the twentieth century. Alfonso would serve on *Independence* until its decommissioning in 1998.

His assignment to the *Independence* saw him much busier than he had been during the last 5 years while stationed in San Diego. Manny was stationed with him at the time and remembered Al at his professional peak during this period.

During one deployment, the ship's evaporator malfunctioned. Without it, there would be no fresh water aboard. The situation was dire. Al rapidly located the broken part for the machine on another ship and had it flown over to the *Independence.* "I think he got a good word from the Chief engineer or some kind of award for that," Manny remarked, pride in his voice.

As a result of his dedication and work effort at sea, Alfonso was awarded two Sea Service Deployment Ribbons, in addition to the Southwest Asia Service and Armed Forces Expeditionary Medals. All were a testament to his service during wartime.

—

After 18 years of service, Alfonso was nearing his 20-year mark, which meant military retirement was on the horizon. On July 9, 1998 he reenlisted for two more years which would have put him safely in the retirement zone. One would guess that he would feel relief and happiness at nearing the end of a successful career. That nothing could stop him from claiming his retirement benefits and returning to San Diego or perhaps the Philippines to raise his children, pick up farming again, and return to a life of peace and stability that always seemed just beyond his grasp during his years in the Navy. But two final obstacles stood in his path.

The first was the anxiety that often assaults career servicemembers as they approach a life without the military. Efren, reflecting on his own struggles transitioning from the military, hypothesized about how retirement might have negatively impacted Alfonso's psychological health.

"...It's the fear of the unknown. I went through it. Because you don't know what's going to happen. Your steady income is going to go away. Yeah. And you have a life. You have a family. You have kids to support. So, I went through that same stress. Normally this [at] the two-year mark. That's when you start attending the transition assistance, retirement program...I can definitely relate to that...You know, you panic. You panic, because you don't know what's going to happen. We were calculating the retirement pay for the military. And it's not that much...We went shopping for a mobile home. And even a mobile home in this area we cannot afford. Wow...So it's not gonna be enough...so it's that fear of the unknown."

Alfonso's concern for his family's financial well-being throughout his career turned the thought of retirement into an unsettling scenario, haunted by shadows of scarcity and anxiety.

The second obstacle was one few were aware of. He wasn't happy.

Manny couldn't remember when or in what context Al said it to him, but during one conversation, in an uncommon moment of emotional vulnerability, Al confided, "I hate my job."

When Manny asked why, Al mentioned dreading work in the "pit," or the guts of the ship. The oven-like closet-sized spaces and messy operation of keeping floating behemoths like the *Independence* powered and operational over the years had stained his psyche, causing him grief so great he had no idea how to communicate it.

Nothing came of the conversation. Alfonso never brought it up again and Manny didn't ask, but he felt something was off about the comment.

He wasn't the only one to note strange behavior in Alfonso. Adelaida (Del) Pamintuan Vandeveer, Alfonso's sister-in-law, remembered an occasion when he had pointed a real gun to his head when he was living in San Diego. She had no idea who the weapon belonged to or if it was loaded, but she knew the unsettling motion scared her.

Was he joking around? Was it silent plea for help? Or was the gesture a telltale signal of deeper darkness within? She had no idea.

CHAPTER 8

Disappearance

---

*"I lost my best buddy...I still charge myself
like I could have done more."*
Melchor Quitoriano
(Alfonso's Friend, Shipmate on
USS Kitty Hawk)

No one knew the depth of Alfonso's depression. Friends, family, and possibly even Alfonso himself, were not aware of its creeping influence on his thoughts and actions. His discreet character coupled with sporadic behaviors he had exhibited throughout his life and career culminated in an event that shocked and mystified all who knew him.

Under normal circumstances disappearing is a difficult thing to do. Typically, someone will see the person somewhere and word will get back to authorities or family members to locate them. But at sea, the situation is much different. When navigating the world's largest and deepest ocean, it is entirely possible for small watercraft to get lost only a few miles from shore. The situation for people is much more merciless. If a hapless person falls into the sea without being seen by someone else or a passing ship, the chances of recovering them is extremely low. A fact that made Alfonso's disappearance that much more heartbreaking.

The events leading up to that fateful day are baffling and piecing them together to construct a coherent picture was a daunting task that required deep examination. What follows is a result of that work and an analysis of the theories surrounding Alfonso's final watch and disappearance.

———

In August 1998, Helen Pamintuan, Efren's wife, received a phone call from Alfonso. He was at a Navy facility near Puget Sound in Washington state and he wanted to speak to her husband, his brother-in-law, Efren. At that time, Efren was at work, so he was unable to take the call. Helen extended an offer for Alfonso to drive to Naval Station Everett, where she and her husband were stationed nearby to share a meal and a place to stay, but Alfonso declined. He was due to return to Japan the following day on the USS Kitty Hawk and would not have enough time to make the drive.

The unexpected nature of the call and the fact Efren never found out why Alfonso contacted him on that day deeply troubled him. The fact that Alfonso had been in the area for a few days and did not contact him until the final day before his departure, bewildered him further. Efren elaborated:

> "So it's one of those [things] if I was able to talk to him, you know, I would [have] encouraged him or...help him. Would [have] met telling him to hang on and things will get better. Could that have made a difference? I don't know.

You know, because again, if somebody is going through what they
call the dark night of the soul,
sometimes they don't hear anything. They don't see anything.
They're in a dark, dark place.
So, you ask for a miracle, you
pray for a miracle that somebody
help you in your situation. Sometimes, there's no help. Sometimes
it's like you're being tested. You
know, sometimes it's like going
through a storm."

That storm manifested metaphorically and physically. On October 10, 1998, a massive mon- soon trough began to gather strength from the churning ocean and atmosphere east of Guam in the Pacific. Three days later, it evolved into Typhoon Zeb, a category 5 super storm that threatened Alfonso's home island of Luzon in the Philippines.

The whirling maelstrom slammed into the island, ripping homes and infrastructure to splinters, leaving millions of dollars in damage in its wake. It then turned north, hitting Taiwan hard before following a straight path toward the main Japanese islands where Alfonso's homeport was located.

Around the same time, the Kitty Hawk departed Yokosuka to prepare for participation in Foal Eagle, the largest joint exercise in the world. It sailed to Busan in South Korea, then headed out to sea to continue its mission, with the typhoon whipping to the west of its path. The ship undoubtedly felt the effects of the storm's gusting winds as it journeyed toward its duty location.

On October 16, 1998 Petty Officer First Class Amos stood watch from 2000 to 0000 hours. His friend, Melchor Quitoriano relieved him at 0000 hours. After he concluded his watch around 0400 hours, he returned to their sleeping quarters, but Alfonso was not there. Quitoriano, dressed in an orange tank-top shirt in his interview, recalled that early morning with sobering clarity:

> "I went down to the birthing compartment and checked [on]

Alfonso. I did not see him. To my suspicion, his blanket was folded tight and neat, and his pillow seemed to me [that] it wasn't even used at all. I did not raise a red flag yet and thought he was roaming around the ship. Then it was time to turn in our daily muster report at 0730 in the morning. That's for all hands. I was in a different workspace, but we shared the same berthing compartment. I learned that he was counted being present when [the] muster report was turned in. Then, silently, the rest of the crew, in his respective work center, started searching for him, and he was nowhere to be found. From A Division, the whole department, the chain of command notified and we went on general quarters (GQ)."

It wasn't long after word reached the command of the Kitty Hawk and a massive search effort began to locate Alfonso. Helicopters from the ship buzzed overhead, combing the windswept waves. Nearby ships immediately joined the manhunt. The USS Chancelorsville, USS Mobile Bay, and US Coast Guard Cutter Jarvis circled the waters with searchlights, binoculars, and hundreds of pairs of eyes angled toward the angry sea below, hoping to catch a glimpse of the missing sailor. The endeavor continued into the night and for another day until 0600 on October 19. It was at that time that the search was terminated.

A casualty report outlined in direct language Alfonso's grim chances for survival. "AT THE TIME OF DISAPPEARANCE, UNIT WAS ON THE OUTSKIRTS OF TYPHOON ZEB, SEA STATES 7 – 12 FEET. GIVEN DISTANCE FROM NEAREST LAND AND SEA STATE, SURVIVAL EXPECTATIONS FOR GREATER THAN 40 HOURS IS CONSIDERED POOR."

Alfonso was declared deceased shortly afterward. His remains were never recovered.

———

The report of casualty issued by the Department of the Navy recorded the cause of death as 'DROWNED.' However, the circumstances were deemed to be 'UNDETERMINED.'

Several mysteries swirl around Alfonso's disappearance. What happened after Quitoriano took over watch duty? Was his death accidental? Or were other nefarious forces at play? To these questions and more, we may never know the answers.

There have been rumors that Alfonso did not perish on that day. Both Manny and Quitoriano claimed to have heard whisperings that placed Alfonso in San Diego or back in Korea during periods since the incident. But none of the sightings were confirmed to be true.

It was also suggested that Alfonso may have accidentally fallen overboard. Dennis Vandeveer, husband of Adelaida (Del) Vandeveer, worked on the flight deck of aircraft carriers while serving in the Navy. The flight deck of an aircraft carrier is a hazardous place. For an unsuspecting sailor, one moment of carelessness could lead to disastrous consequences. Dennis had heard stories of crew being blown off deck by aircraft jet blast, high winds or even slipping over the side due to rough seas. When considering Alfonso's disappearance, he commented:

> "If they go over during the day, then there's a good chance of rescue. Cause you've got other ships in the area, helicopters out...At night, even if they know they just fell over the side, and that fast it's really really hard to find them, and if they don't have the safety equipment on; the vest and the light...then it's really tough to find em.' Cause it's dark, the ocean's dark."

He thought it was likely Alfonso went above deck to cool down for relief from the blistering temperatures below decks, then was washed overboard due to the violent seas caused by the typhoon. While this is certainly a plausible explanation, no evidence or witnesses exist to corroborate it.

———

The last person to see Alfonso alive was Quitoriano. During their shift change they had one last conversation. Quitoriano remembered Alfonso seemed noticeably quiet and said few words. "He was quiet when we left Busan," Quitoriano said, "like he was hiding something." He wondered if Alfonso had an argument with his wife or if something else had happened. If such an event had occurred Alfonso provided no details.

After the search concluded, Quitoriano was interviewed by several members of the ship's command as a person of interest. He described himself as "devastated" by the shocking news of his friend's disappearance.

> "I felt guilty. I keep asking myself, what could I have done more to [stop] this thing to happen [sic]. Or, what did I do wrong?...I still

charge myself I could have done more."

During the final moments of his interview for this book, Quitoriano sat up straight and with a voice wavering with emotion said,

> "We were good friends...I could call [him] as a friend when [he] was willing to give his life for me. That's the kind of friendship we had...Losing a friend...it's a family thing...and I hope this will serve for you (Josephine), for the family, for myself too, as a good way of reflecting, remembering what your dad have [sic] done for his family, for you. I'm not talking about for himself. You know why? That's the kind of person he was, he was a selfless person...a good sailor...a good machinist mate."

CHAPTER 9

Seeking Peace

"The whole process is kind of a lot of emotional labor but at the same time it's been so beautiful having all these interviews. Yesterday was so touching. And just seeing that weight lift off from my dad's friends and just having that space to talk about it, is so priceless and so important...I'm getting a more clear [sic] picture of who he was and the impact and the people who loved him."

Josephine Pamintuan Amos
(Alfonso's Youngest Daughter)

Josephine Amos was only 5-years-old when her father disappeared. Without the emotional capacity to truly process his absence in her life, she didn't think much about it until she became older.

> "In my pre-teens or early high school, I remember really thinking about how it affected my mom and just really understanding her being a widow, a single mom, and raising us. That's when it first hit me...It didn't really hurt losing my dad, but the grief of what my mom went through really got to me. I don't know what happened, but I just remember being really sad about it in my early teens and then in college."

When she left home to attend college, she felt distant from her family. "It didn't really feel warm. It didn't feel cohesive. It didn't feel like they understood me or knew me and at the same time I didn't know them."

As she began to explore her identity throughout her college years, she became more interested in discovering her cultural roots, her family history, and eventually what happened to her father.

She described the moment during a spiritual retreat that inspired her to formally investigate his life and death.

> "One of the things that we did was this long meditation. And when everyone came out of it, they shared what they experienced and a lot of it was about connecting with our ancestors and connecting to who we are. I just remember everyone sharing and I didn't personally experience anything that profound during my meditation, but in that moment of listening to everyone's share, I suddenly like felt this black hole. It felt like this deep lack of knowing who I was and I connected it to not knowing my father. And I just remember sitting in that circle crying my eyes out, like '*Oh my God, like I need to know*

> *him, and like I can't believe I don't know him'*... so in that moment I was like, *'I'm gonna investigate dad's death'.*"

Josephine, now a young woman with dark wavy hair and the soul of a seeker, began her investigation. She tracked down and interviewed family members scattered around the United States and Philippines, found, and analyzed key documents from her dad's Navy career, and, with her sister Jennifer's help, coordinated with The Price of Freedom Foundation that culminated in this work. Her father would surely be proud of the intelligent and self-sufficient woman she had become and her great effort in preserving his life and legacy.

True to her introspective nature, Josephine shared her thoughts with respect to what she learned from the process.

> "I just wish that he had more support and felt more comfortable opening up to people. But as far as I've heard he was very loved

and liked and a good person and a hard worker and very intelligent. He loved naps and he loved food. And so just learning all these things about him has been very fulfilling."

———

Alfonso sits before an American flag in his full dress blue US Navy uniform. His hair is neatly groomed. His left arm is angled toward the camera. A sewn on patch with a white eagle with wings spread wide over a ship propeller of the same color above three red colored chevrons pointing downward form his rank, Machinist Mate 1st Class. His expression is non-descript, but the lips are slightly parted, as if he had more he wanted to say, but never had the opportunity to.

Military official photos give all service-members a veil of invincibility. Seated in front of an American flag that appears to flutter in the background, a row of medals or awards on the chest and dressed to perfection, it's easy to forget the heart beating behind the military decorations is not that of a soulless robot who executes orders, but of a person with hopes, dreams, and fears.

Alfonso Apdal Amos was a family man with simple desires.

He loved farming, fishing, listening to music on his Walkman cassette player, and most importantly spending time with his family. Although quiet by nature, his silent soul expressed itself through his caring and action. A trait his family and friends found endearing and charming.

The legacy of his life lives on. Not just through this work, but also through the lives of his children and his military service. From the jungles of Mindanao to the deepest oceans, he sacrificed substantial portions of his life for his family and both nations he called home.

Let his life serve as a reminder to check on fellow veterans in times of stress. Let no one remain a silent witness or allow a fellow servicemember or veteran to bear a load that may be too great a burden. One loss to suicide is too many for the military and for loved ones.

"Sometimes people ask, why suffering? What's the point of suffering? And that's an age-old question. Because everybody's goal in life is to be happy, to be joyful. So there's good that comes out of suffering. Look at your family. Your mom has gone through that tragic experience and recovered from it. And you guys, you (Josephine), your dad's dream about all of you to finish school, that happened. You know, he suffered, he sacrificed for you guys. And all of you completed your college degrees. So, going through that tragedy, that experience the good that came out of that, is your family. You know, you're all successful as far as education, and work. So that was his dream...and now you're here. So, if he's alive, I'm sure he's gonna celebrate what happened to [his] kids. So that's the second thing that came out of that is there is resiliency, yeah. That in spite of what the family went through, you guys persevered."

-Francisco (Efren) Pamintuan (Alfonso's brother-in-law)

Bibliography

Preface

Typhoon Zeb - Wikipedia, Wikipedia Foundation. Retrieved 21 Mar 2022.

Digital Typhoon: Typhoon 199810 (ZEB) - General Information (Pressure and Track Charts) (nii.ac.jp), Retrieved 21 Mar 2022.

Chapter 2: Armed Rebels

Ahmad, Aijaz (1982) Class and Colony in Mindanao. *Southeast Asia Chronicle* No. 82:4-11.

Email. "Email from Jen's Scholar Friend," From Jennifer Amos. 8 Oct 2021

George, T. J. S. (1980). *Revolt in Mindanao: The Rise of Islam in Philippine Politics.* Oxford University Press. pp. 130–134.

Kamilan, Jamail A (2012). "Who are the Moro people?," Retrieved 4 Nov 2021

Majul, Cesar Adlib (1981). "An Analysis of the "Genealogy of Sulu"".

Majul, Cesar Adib (1985) The Contemporary Muslim Movement in the Philippines. Berkeley: Mizan Press.

McKenna, Thomas M. (1998) Muslim Rulers and Rebels: Everyday Politics and Armed Separatism in the Southern Philippines. Berkeley: University of California Press.

Reyes, Rachel A.G. (April 16, 2016). "3,257: Fact checking the Marcos killings, 1975-1985 - The Manila Times Online". *Manila Times*. Retrieved June 15, 2018.

Robles, Raissa (2016). *Marcos Martial Law: Never Again*. FILIPINOS FOR A BETTER PHILIPPINES, INC.

Smith, Paul J. (March 26, 2015). *Terrorism and Violence in Southeast Asia: Transnational Challenges to States and Regional Stability: Transnational*

Challenges to States and Regional Stability. Taylor & Francis.

Philippine Claim to North Borneo (Sabah), Vol. II. Government of the Philippines. August 2, 1962. "I. North Borneo Claim (Excerpt from President Diosdado Macapagal's State-of-the-Nation Message to the Congress of the Philippines)". *Philippine Claim to North Borneo (Sabah), Vol. II.* Government of the Philippines. January 28, 1963.

Chapter 3: American Sailor

https://en.wikipedia.org/wiki/Naval_Training_Center_San_Diego#cite_note-Quarterdeck:_NTC_during_the_1960s-9, Wikipedia Foundation. Retrieved 10 Nov 2021

https://phc.amedd.army.mil/PHC%20Resource%20Library/Barracks%20Layout%20Jan%202010.pdf *Just The Facts – Barracks Layout to Prevent Disease Transmission* 36-017-0110 U.S. Army Public Health Command (Provisional). Retrieved 11 July 2022

https://weatherspark.com/h/m/1816/1980/7/Historical-Weather-in-July-1980-in-San-Diego-California-United-States#Figures-Temperature, WeatherSpark. Retrieved 10 Nov 2021.

https://www.operationmilitarykids.org/navy-machinists-mate-mm/, Operation MilitaryKids. Retrieved 11 Nov 2021

https://www.mynavyhr.navy.mil/References/Instructions/BUPERS-Instructions/, MyNavyHR. Retrieved 11 Nov 2021.

USA 20 mm Phalanx Close-in Weapon System (CIWS) - NavWeaps, NavWeaps. Retrieved 22 Apr 2022.

Chapter 4: A Man Defined

https://historycentral.com/navy/CVB41Midway.html. Retrieved 15 Nov 2021

https://www.midway.org/about-us/midway-history/ Retrieved 18 Nov 2021

Prince, Troy. "The History of Midway's Magic." https://www.midwaysailor.com/midway/history.html. Retrieved 18 Nov 2021

Olongapo - Wikipedia. Wikipedia Foundation. Retrieved 21 Nov 2021

https://panlasangpinoy.com/papaitan-recip/, Panlasang Pinoy. Retrieved 29 Nov 2021.

Chapter 6: Going Through Darkness

Wise, Harold Lee (2007). *Inside the Danger Zone: The U.S. Military in the Persian Gulf 1987–88*. Annapolis: Naval Institute Press.

McCulloch, Cameron (2014) What it takes to prepare a US Navy ship for deployment: Kicking the tires, lighting the fires DVIDS - News - What it takes to prepare a US Navy ship for deployment: Kicking the tires, lighting the fires (dvidshub.net)

Chapter 7: Career of a Machinist Mate

USS ENTERPRISE (CVN-65) Deployments & History (hullnumber.com). Hullnumber.com. Retrieved 16 Apr 2022.

USS INDEPENDENCE (CV-62) Deployments & History (hullnumber.com). Hullnumber.com. Retrieved 16 Apr 2022.

USS Kitty Hawk (CV 63) history (uscarriers.net). Hullnumber.com. Retrieved 20 Apr 2022.

Prince, Troy. "The History of Midway's Magic." https://www.midwaysailor.com/midway/history.html. Retrieved 16 Apr 2022

Wilson, George C (1983). "Enterprise Gets Stuck on San Francisco Bar," The Washington Post. Enterprise Gets Stuck On San Francisco Bar - The Washington Post

www.ingramcontent.com/pod-product-compliance
Lightning Source LLC
Chambersburg PA
CBHW051627120626
46551CB00014B/1962